Eureka Math™
A Story of Units

Special thanks go to the Gordan A. Cain Center and to the Department of Mathematics at Louisiana State University for their support in the development of Eureka Math.

Published by Common Core

Copyright © 2014 Common Core, Inc. All rights reserved. No part of this work may be reproduced or used in any form or by any means – graphic, electronic, or mechanical, including photocopying or information storage and retrieval systems – without written permission from the copyright holder. "Common Core" and "Common Core, Inc.," are registered trademarks of Common Core, Inc.

Common Core, Inc. is not affiliated with the Common Core State Standards Initiative.

Printed in the U.S.A.
This book may be purchased from the publisher at commoncore.org
10 9 8 7 6 5 4 3 2

ISBN 978-1-63255-031-6

Name _____   Date _____

1. Shade the first 7 units of the tape diagram. Count by tenths to label the number line using a fraction and a decimal for each point. Circle the decimal that represents the shaded part.

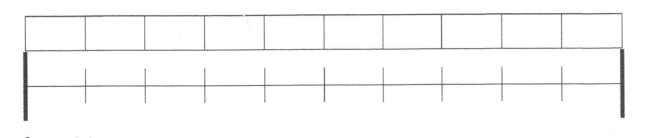

0   0.1   ___   ___   ___   ___   ___   ___   ___   ___   1

       $\frac{1}{10}$

2. Write the total amount of water in fraction form and decimal form. Shade the last bottle to show the correct amount.

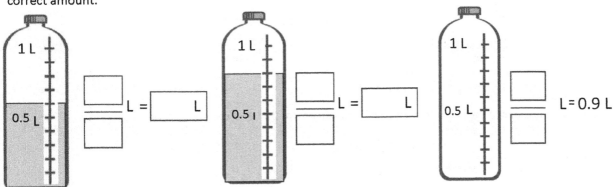

3. Write the total weight of the food on each scale in fraction form or decimal form.

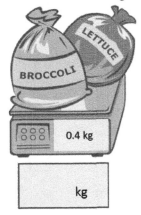

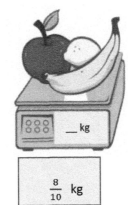

0.4 kg

_____ kg

kg

$\frac{8}{10}$ kg

kg

EUREKA MATH

Lesson 1:   Use metric measurement to model the decomposition of one whole into tenths.

1

4. Write the length of the bug in centimeters. (Drawing is not to scale.)

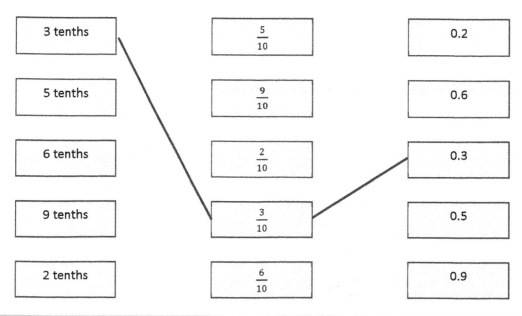

Fraction form: _____ cm

Decimal form: _____ cm

How far does the bug need to walk before its nose is at the 1 cm mark? _____ cm

5. Fill in the blank to make the sentence true in both fraction form and decimal form.

a. $\frac{8}{10}$ cm + _____ cm = 1 cm   0.8 cm + _____ cm = 1.0 cm

b. $\frac{2}{10}$ cm + _____ cm = 1 cm   0.2 cm + _____ cm = 1.0 cm

c. $\frac{6}{10}$ cm + _____ cm = 1 cm   0.6 cm + _____ cm = 1.0 cm

6. Match each amount expressed in unit form to its equivalent fraction and decimal forms.

3 tenths		$\frac{5}{10}$		0.2
5 tenths		$\frac{9}{10}$		0.6
6 tenths		$\frac{2}{10}$		0.3
9 tenths		$\frac{3}{10}$		0.5
2 tenths		$\frac{6}{10}$		0.9

Name _____   Date _____

. Shade the first 4 units of the tape diagram. Count by tenths to label the number line using a fraction and a decimal for each point. Circle the decimal that represents the shaded part.

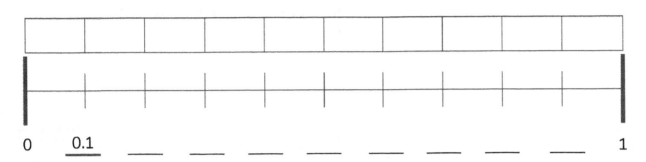

0   0.1   ___   ___   ___   ___   ___   ___   ___   ___   ___   1

         $\frac{1}{10}$

2. Write the total amount of water in fraction form and decimal form. Shade the last bottle to show the correct amount.

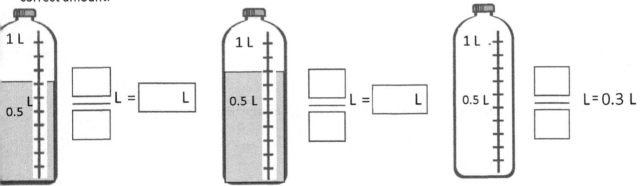

3. Write the total weight of the food on each scale in fraction form or decimal form.

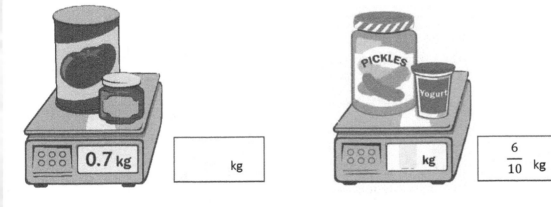

0.7 kg   ____ kg   ____ kg $\frac{6}{10}$ kg

EUREKA
MATH™

Lesson 1:   Use metric measurement to model the decomposition of one whole
             into tenths.

3

4. Write the length of the bug in centimeters. (Drawing is not to scale.)

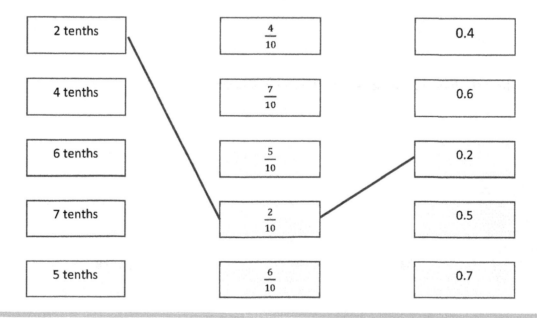

Fraction form: _____ cm

Decimal form: _____ cm

If the bug walks 0.5 cm farther, where will its nose be? _____ cm

5. Fill in the blank to make the sentence true in both fraction and decimal form.

a. $\frac{4}{10}$ cm + _____ cm = 1 cm   0.4 cm + _____ cm = 1.0 cm

b. $\frac{3}{10}$ cm + _____ cm = 1 cm   0.3 cm + _____ cm = 1.0 cm

c. $\frac{8}{10}$ cm + _____ cm = 1 cm   0.8 cm + _____ cm = 1.0 cm

6. Match each amount expressed in unit form to its equivalent fraction and decimal.

2 tenths		$\frac{4}{10}$		0.4
4 tenths		$\frac{7}{10}$		0.6
6 tenths		$\frac{5}{10}$		0.2
7 tenths		$\frac{2}{10}$		0.5
5 tenths		$\frac{6}{10}$		0.7

Name _____   Date _____

1. For each length given below, draw a line segment to match. Express each measurement as an equivalent mixed number.

a. 2.6 cm

b. 3.4 cm

c. 3.7 cm

d. 4.2 cm

e. 2.5 cm

2. Write the following as equivalent decimals. Then, model and rename the number as shown below.

a. 2 ones and 6 tenths = _____

 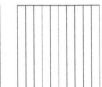 

$2\frac{6}{10} = 2 + \frac{6}{10} = 2 + 0.6 = 2.6$

EUREKA MATH | Lesson 2:   Use metric measurement and area models to represent tenths as fractions greater than 1 and decimal numbers.

5

b. 4 ones and 2 tenths = _____

c. $3\frac{4}{10}$ = _____

d. $2\frac{5}{10}$ = _____

How much more is needed to get to 5? _____

e. $\frac{37}{10}$ = _____

How much more is needed to get to 5? _____

EUREKA MATH™

Lesson 2:   Use metric measurement and area models to represent tenths as
fractions greater than 1 and decimal numbers.

Name _____   Date _____

1. For each length given below, draw a line segment to match. Express each measurement as an equivalent mixed number.

   a.   2.6 cm

   b.   3.5 cm

   c.   1.7 cm

   d.   4.3 cm

   e.   2.2 cm

2. Write the following in decimal form. Then, model and rename the number as shown below.

   a.   2 ones and 4 tenths = _____

$$2\frac{4}{10} = 2 + \frac{4}{10} = 2 + 0.4 = 2.4$$

EUREKA MATH™

Lesson 2:   Use metric measurement and area models to represent tenths as fractions greater than 1 and decimal numbers.

7

b. 3 ones and 8 tenths = _____

c. $4\frac{1}{10}$ = _____

d. $1\frac{4}{10}$ = _____

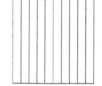  

How much more is needed to get to 5? _____

e. $\frac{33}{10}$ = _____

  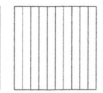 

How much more is needed to get to 5? _____

EUREKA MATH™

Lesson 2: Use metric measurement and area models to represent tenths as fractions greater than 1 and decimal numbers.

tenths area model

Lesson 2:   Use metric measurement and area models to represent tenths as
            fractions greater than 1 and decimal numbers.

9

Name _____   Date _____

. Circle groups of tenths to make as many ones as possible.

a. How many tenths in all?	Write and draw the same number using ones and tenths.
    There are _____ tenths.	 Decimal Form: _____ How much more is needed to get to 3? _____
b. How many tenths in all?	Write and draw the same number using ones and tenths.
  There are _____ tenths.	 Decimal Form: _____ How much more is needed to get to 4? _____

2. Draw disks to represent each number using tens, ones, and tenths. Then, show the expanded form of the number in fraction form and decimal form as shown. The first one has been completed for you.

a. 4 tens 2 ones 6 tenths	b. 1 ten 7 ones 5 tenths
  Fraction Expanded Form $(4 \times 10) + (2 \times 1) + (6 \times \frac{1}{10}) = 42\frac{6}{10}$ Decimal Expanded Form $(4 \times 10) + (2 \times 1) + (6 \times 0.1) = 42.6$	

c.	2 tens 3 ones 2 tenths	d.	7 tens 4 ones 7 tenths

3. Complete the chart.

Point	Number Line	Decimal Form	Mixed Number (ones and fraction form)	Expanded Form (fraction or decimal form)	How much to get to the next one?
a.			$3\frac{9}{10}$		0.1
b.	17 18				
c.				$(7 \times 10) + (4 \times 1) + (7 \times \frac{1}{10})$	
d.			$22\frac{2}{10}$		
e.				$(8 \times 10) + (8 \times 0.1)$	

Name _____   Date _____

1. Circle groups of tenths to make as many ones as possible.

a. How many tenths in all?  0.1 0.1 0.1 0.1 0.1 0.1 0.1 0.1 0.1 0.1 0.1 0.1 0.1 0.1 There are _____ tenths.	Write and draw the same number using ones and tenths. Decimal Form: _____ How much more is needed to get to 2? _____
b. How many tenths in all? 0.1 0.1 0.1 0.1 0.1 0.1 0.1 0.1 0.1 0.1 0.1 0.1 0.1 0.1 0.1 0.1 0.1 0.1 0.1 0.1 0.1 0.1 0.1 0.1 0.1 There are _____ tenths.	Write and draw the same number using ones and tenths. Decimal Form: _____ How much more is needed to get to 3? _____

2. Draw disks to represent each number using tens, ones, and tenths. Then, show the expanded form of the number in fraction form and decimal form as shown. The first one has been completed for you.

a. 3 tens 4 ones 3 tenths	b. 5 tens 3 ones 7 tenths
 10 10 10 1 1 1 1 0.1 0.1 0.1 Fraction Expanded Form $(3 \times 10) + (4 \times 1) + (3 \times \frac{1}{10}) = 34\frac{3}{10}$ Decimal Expanded Form $(3 \times 10) + (4 \times 1) + (3 \times 0.1) = 34.3$	

EUREKA MATH™

Lesson 3: | Represent mixed numbers with units of tens, ones, and tenths with number disks, on the number line, and in expanded form.

13

c. 3 tens 2 ones 3 tenths	d. 8 tens 4 ones 8 tenths

3. Complete the chart.

Point	Number Line	Decimal Form	Mixed Number (ones and fraction form)	Expanded Form (fraction or decimal form)	How much to get to the next one?
a.			$4\frac{6}{10}$		
b.					0.5
c.				$(6 \times 10) + (3 \times 1) + (6 \times \frac{1}{10})$	
d.			$71\frac{3}{10}$		
e.				$(9 \times 10) + (9 \times 0.1)$	

Lesson 3:   Represent mixed numbers with units of tens, ones, and tenths with number disks, on the number line, and in expanded form.

Point	Number Line	Decimal Form	Mixed Number (ones and fraction form)	Expanded Form (fraction or decimal form)	How much more is needed to get to the next one?
a.					
b.					
c.					
d.					

tenths on a number line

Lesson 3: Represent mixed numbers with units of tens, ones, and tenths with number disks, on the number line, and in expanded form.

Name _____ Date _____

a. What is the length of the shaded part of the meter stick in centimeters?

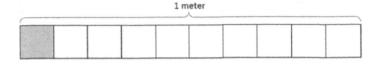

b. What fraction of a meter is 1 centimeter?

c. In fraction form, express the length of the shaded portion of the meter stick.

d. In decimal form, express the length of the shaded portion of the meter stick.

e. What fraction of a meter is 10 centimeters?

2. Fill in the blanks.

a. 1 tenth = _____ hundredths

b. $\frac{1}{10}$ m = $\frac{}{100}$ m

c. $\frac{2}{10}$ m = $\frac{20}{}$ m

3. Use the model to add the shaded parts as shown. Write a number bond with the total written in decimal form and the parts written as fractions. The first one has been done for you.

a.

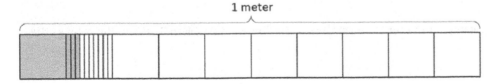

$$\frac{1}{10}\text{ m} + \frac{3}{100}\text{ m} = \frac{13}{100}\text{ m} = 0.13\text{ m}$$

EUREKA MATH™

Lesson 4: Use meters to model the decomposition of one whole into hundredths. Represent and count hundredths.

17

b.

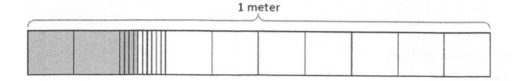

c.

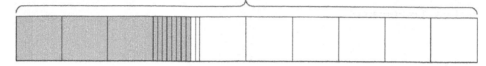

4. On each meter stick, shade in the amount shown. Then, write the equivalent decimal.

a. $\frac{8}{10}$ m

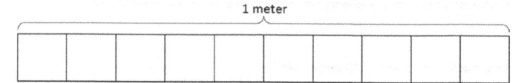

b. $\frac{7}{100}$ m

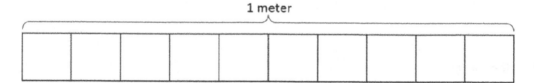

c. $\frac{19}{100}$ m

5. Draw a number bond, pulling out the tenths from the hundredths as in Problem 3. Write the total as the equivalent decimal.

a. $\frac{19}{100}$ m

b. $\frac{28}{100}$ m

c. $\frac{77}{100}$

d. $\frac{94}{100}$

Name _____   Date _____

1. a. What is the length of the shaded part
        of the meter stick in centimeters?

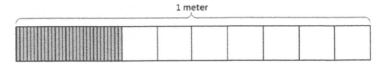

    b. What fraction of a meter is 3 centimeters?

    c. In fraction form, express the length of
        the shaded portion of the meter stick.

    d. In decimal form, express the length of the shaded portion of the meter stick.

    e. What fraction of a meter is 30 centimeters?

2. Fill in the blanks.

    a. 5 tenths = _____ hundredths

    b. $\frac{5}{10}$ m = $\frac{}{100}$ m

    c. $\frac{4}{10}$ m = $\frac{40}{}$ m

3. Use the model to add the shaded parts as shown. Write a number bond with the total written in decimal
    form and the parts written as fractions. The first one has been done for you.

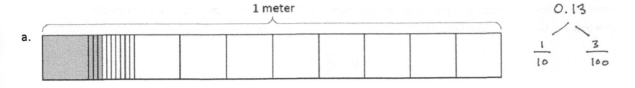

    a.

    $\frac{1}{10}$ m + $\frac{3}{100}$ m = $\frac{13}{100}$ m = 0.13 m

EUREKA
MATH™

Lesson 4:   Use meters to model the decomposition of one whole into hundredths.
            Represent and count hundredths.

19

b.

1 meter

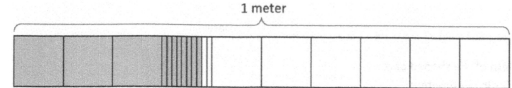

c.

1 meter

4. On each meter stick, shade in the amount shown. Then, write the equivalent decimal.

1 meter

a. $\frac{9}{10}$ m

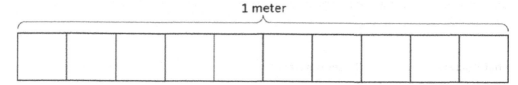

1 meter

b. $\frac{15}{100}$ m

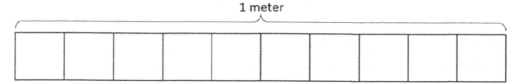

1 meter

c. $\frac{41}{100}$ m

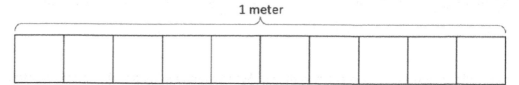

5. Draw a number bond, pulling out the tenths from the hundredths, as in Problem 3 of the Homework.
    Write the total as the equivalent decimal.

a. $\frac{23}{100}$ m

b. $\frac{38}{100}$ m

c. $\frac{82}{100}$

d. $\frac{76}{100}$

EUREKA
MATH™ | Lesson 4:   Use meters to model the decomposition of one whole into hundredths.
                                    Represent and count hundredths.

1 meter

1 meter

1 meter

1 meter

1 meter

tape diagram in tenths

Lesson 4: Use meters to model the decomposition of one whole into hundredths.
Represent and count hundredths.

ame _____   Date _____

1. Find the equivalent fraction using multiplication or division. Shade the area models to show the equivalency. Record it as a decimal.

a. $\dfrac{3 \times}{10 \times} = \dfrac{}{100}$

 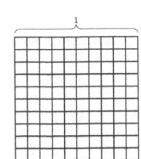

b. $\dfrac{50 \div}{100 \div} = \dfrac{}{10}$

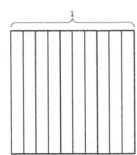

2. Complete the number sentences. Shade the equivalent amount on the area model, drawing horizontal lines to make hundredths.

a. 37 hundredths = _____ tenths + _____ hundredths

   Fraction form: _____

   Decimal form: _____

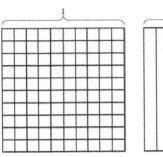

b. 75 hundredths = _____ tenths + _____ hundredths

   Fraction form: _____

   Decimal form: _____

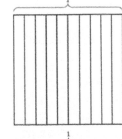

3. Circle hundredths to compose as many tenths as you can. Complete the number sentences. Represent each with a number bond as shown.

a.

_____ hundredths = _____ tenth + _____ hundredths

EUREKA
MATH™

Lesson 5: Model the equivalence of tenths and hundredths using the area model and number disks.

23

b.

(0.01) (0.01) (0.01) (0.01) (0.01)   (0.01) (0.01) (0.01) (0.01) (0.01)

(0.01) (0.01) (0.01) (0.01) (0.01)   (0.01) (0.01)

(0.01) (0.01) (0.01) (0.01) (0.01)

(0.01) (0.01) (0.01) (0.01) (0.01)   _____ hundredths = _____ tenths + _____ hundredths

4. Use both tenths and hundredths number disks to represent each number. Write the equivalent number in decimal, fraction, and unit form.

a. $\frac{3}{100}$ = 0. _____ _____ hundredths	b. $\frac{15}{100}$ = 0. _____ _____ tenth _____ hundredths
c. —— = 0.72 _____ hundredths	d. —— = 0.80 _____ tenths
e. —— = 0. _____ 7 tenths 2 hundredths	f. —— = 0. _____ 80 hundredths

Name _____   Date _____

Find the equivalent fraction using multiplication or division. Shade the area models to show the equivalency. Record it as a decimal.

a. $\dfrac{4 \times \underline{}}{10 \times \underline{}} = \dfrac{\underline{}}{100}$

b. $\dfrac{60 \div \underline{}}{100 \div \underline{}} = \dfrac{\underline{}}{10}$

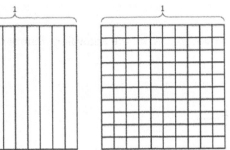

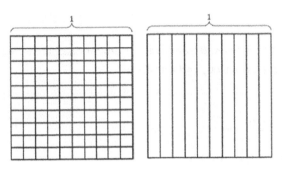

. Complete the number sentences. Shade the equivalent amount on the area model, drawing horizontal lines to make hundredths.

a. 36 hundredths = _____ tenths + _____ hundredths

    Decimal form: _____

    Fraction form: _____

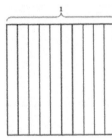

b. 82 hundredths = _____ tenths + _____ hundredths

    Decimal form: _____

    Fraction form: _____

. Circle hundredths to compose as many tenths as you can. Complete the number sentences. Represent each with a number bond as shown.

a.

0.14

$\dfrac{1}{10}$ $\dfrac{4}{100}$

_____ hundredths = _____ tenth + _____ hundredths

Lesson 5:   Model the equivalence of tenths and hundredths using the area model and number disks.

b. (0.01)(0.01)(0.01)(0.01)(0.01) (0.01)(0.01)(0.01)(0.01)

(0.01)(0.01)(0.01)(0.01)(0.01)

(0.01)(0.01)(0.01)(0.01)(0.01)

(0.01)(0.01)(0.01)(0.01)(0.01)   _____ hundredths = _____ tenths + _____ hundredths

4. Use both tenths and hundredths number disks to represent each number. Write the equivalent number in decimal, fraction, and unit form.

a. $\frac{4}{100} = 0.$ _____ _____ hundredths	b. $\frac{13}{100} = 0.$ _____ _____tenth _____ hundredths
c. —— = 0.41 _____ hundredths	d. —— = 0.90 _____ tenths
e. —— = 0. _____ 6 tenths 3 hundredths	f. —— = 0. _____ 90 hundredths

EUREKA MATH™

Lesson 5: Model the equivalence of tenths and hundredths using the area model and number disks.

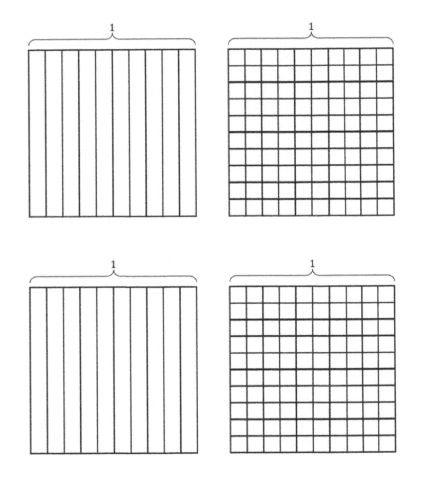

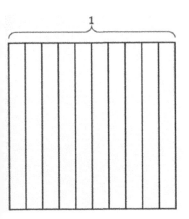
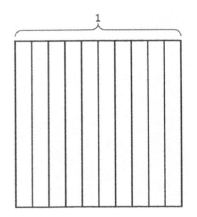
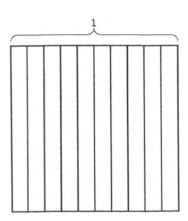

tenths and hundredths area model

EUREKA
MATH™

Lesson 5:   Model the equivalence of tenths and hundredths using the area model
             and number disks.

27

Name _____   Date _____

1. Shade the area models to represent the number, drawing horizontal lines to make hundredths as needed. Locate the corresponding point on the number line. Label with a point, and record the mixed number as a decimal.

a. $1\frac{15}{100}$ = ___.___

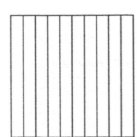 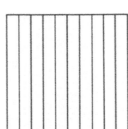 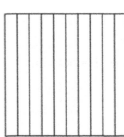

0   1   2   3

b. $2\frac{47}{100}$ = ___.___

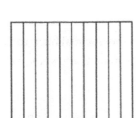 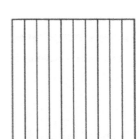

2   3

2. Estimate to locate the points on the number lines.

a. $2\frac{95}{100}$ b. $7\frac{52}{100}$

   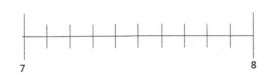

2   3   7   8

EUREKA
MATH™

Lesson 6:   Use the area model and number line to represent mixed numbers with units of ones, tenths, and hundredths in fraction and decimal forms.

29

3. Write the equivalent fraction and decimal for each of the following numbers.

a. 1 one 2 hundredths	b. 1 one 17 hundredths
c. 2 ones 8 hundredths	d. 2 ones 27 hundredths
e. 4 ones 58 hundredths	f. 7 ones 70 hundredths

4. Draw lines from dot to dot to match the decimal form to both the unit form and fraction form. All unit forms and fractions have at least one match, and some have more than one match.

7 ones 13 hundredths ●   ● 7.30 ●   ● $7\frac{3}{100}$

7 ones 3 hundredths ●   ● 7.3 ●   ● 73

7 ones 3 tenths ●   ● 7.03 ●   ● $7\frac{13}{100}$

7 tens 3 ones ●   ● 7.13 ●   ● $7\frac{30}{100}$

          ● 73 ●

EUREKA MATH™

Lesson 6:   Use the area model and number line to represent mixed numbers with units of ones, tenths, and hundredths in fraction and decimal forms.

Name _____ Date _____

1. Shade the area models to represent the number, drawing horizontal lines to make hundredths as needed. Locate the corresponding point on the number line. Label with a point, and record the mixed number as a decimal.

a. $2\frac{35}{100}$ = ___.___

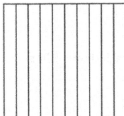 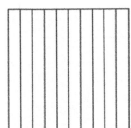 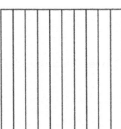

0   1   2   3

b. $3\frac{17}{100}$ = ___.___

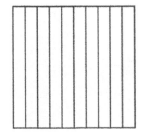 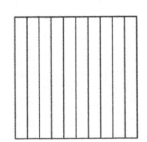 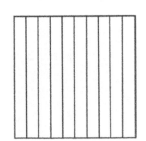 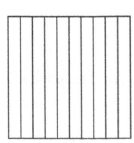

3   4

2. Estimate to locate the points on the number lines.

a. $5\frac{90}{100}$

b. $3\frac{25}{100}$

5   6   3   4

EUREKA MATH™

Lesson 6: Use the area model and number line to represent mixed numbers with units of ones, tenths, and hundredths in fraction and decimal forms.

31

3. Write the equivalent fraction and decimal for each of the following numbers.

a. 2 ones 2 hundredths	b. 2 ones 16 hundredths
c. 3 ones 7 hundredths	d. 1 one 18 hundredths
e. 9 ones 62 hundredths	f. 6 ones 20 hundredths

4. Draw lines from dot to dot to match the decimal form to both the unit form and fraction form. All unit forms and fractions have at least one match, and some have more than one match.

	● 4.80 ●		● $4\frac{18}{100}$
4 ones 18 hundredths ●			
	● 4.8 ●		● 48
4 ones 8 hundredths ●			
	● 4.18 ●		● $4\frac{8}{100}$
4 ones 8 tenths ●			
	● 4.08 ●		● $4\frac{80}{100}$
4 tens 8 ones ●			
	● 48 ●		

EUREKA MATH™

Lesson 6: Use the area model and number line to represent mixed numbers with units of ones, tenths, and hundredths in fraction and decimal forms.

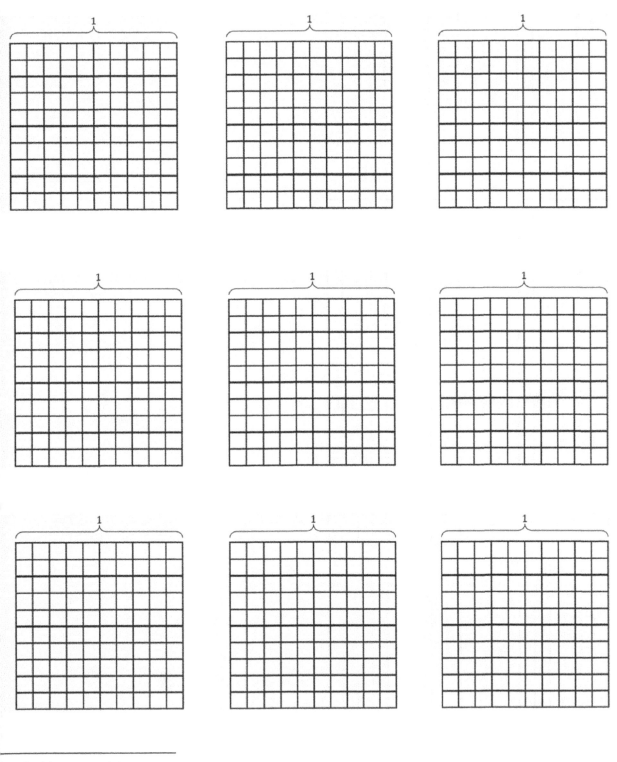

hundredths area model

Lesson 6:   Use the area model and number line to represent mixed numbers with units of ones, tenths, and hundredths in fraction and decimal forms.

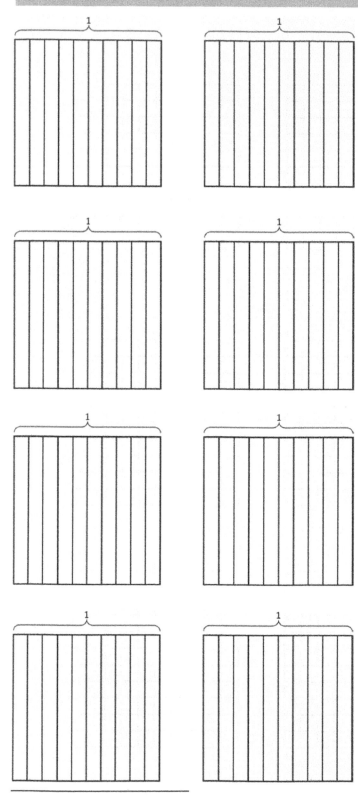

area model

Lesson 6:   Use the area model and number line to represent mixed numbers with units of ones, tenths, and hundredths in fraction and decimal forms.

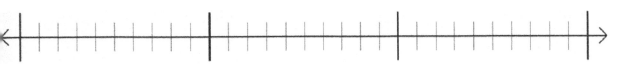

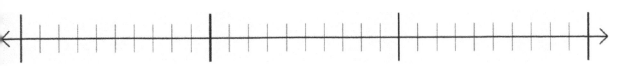

number line

Lesson 6:   Use the area model and number line to represent mixed numbers with units of ones, tenths, and hundredths in fraction and decimal forms.

Name _____   Date _____

1. Write a decimal number sentence to identify the total value of the number disks.

a.

    2 tens   5 tenths   3 hundredths

_____ + _____ + _____ = _____

b.

    5 hundreds   4 hundredths

_____ + _____ = _____

2. Use the place value chart to answer the following questions. Express the value of the digit in unit form.

hundreds	tens	ones	.	tenths	hundredths
4	1	6		8	3

a. The digit _____ is in the hundreds place. It has a value of _____.

b. The digit _____ is in the tens place. It has a value of _____.

c. The digit _____ is in the tenths place. It has a value of _____.

d. The digit _____ is in the hundredths place. It has a value of _____.

hundreds	tens	ones	.	tenths	hundredths
5	3	2		1	6

e. The digit _____ is in the hundreds place. It has a value of _____.

f. The digit _____ is in the tens place. It has a value of _____.

g. The digit _____ is in the tenths place. It has a value of _____.

h. The digit _____ is in the hundredths place. It has a value of _____.

3. Write each decimal as an equivalent fraction. Then, write each number in expanded form, using both decimal and fraction notation. The first one has been done for you.

Decimal and Fraction Form	Expanded Form	
	Fraction Notation	Decimal Notation
$15.43 = 15\frac{43}{100}$	$(1 \times 10) + (5 \times 1) + (4 \times \frac{1}{10}) + (3 \times \frac{1}{100})$ $10 \quad + \quad 5 \quad + \quad \frac{4}{10} \quad + \quad \frac{3}{100}$	$(1 \times 10) + (5 \times 1) + (4 \times 0.1) + (3 \times 0.01)$ $10 \quad + \quad 5 \quad + \quad 0.4 \quad + \quad 0.03$
$21.4 = \underline{\hspace{1cm}}$		
$38.09 = \underline{\hspace{1cm}}$		
$50.2 = \underline{\hspace{1cm}}$		
$301.07 = \underline{\hspace{1cm}}$		
$620.80 = \underline{\hspace{1cm}}$		
$800.08 = \underline{\hspace{1cm}}$		

EUREKA MATH™

Lesson 7:   Model mixed numbers with units of hundreds, tens, ones, tenths, and hundredths in expanded form and on the place value chart.

Name _____   Date _____

. Write a decimal number sentence to identify the total value of the number disks.

a.

(10) (10) (10)   (0.1) (0.1) (0.1) (0.1)   (0.01) (0.01)

    3 tens   4 tenths   2 hundredths

_____ + _____ + _____ = _____

b.

(100) (100) (100) (100)   (0.01) (0.01) (0.01)

   4 hundreds   3 hundredths

_____ + _____ = _____

. Use the place value chart to answer the following questions. Express the value of the digit in unit form.

hundreds	tens	ones	.	tenths	hundredths
8	2	7		6	4

a. The digit _____ is in the hundreds place. It has a value of _____.

b. The digit _____ is in the tens place. It has a value of _____.

c. The digit _____ is in the tenths place. It has a value of _____.

d. The digit _____ is in the hundredths place. It has a value of _____.

hundreds	tens	ones	.	tenths	hundredths
3	4	5		1	9

e. The digit _____ is in the hundreds place. It has a value of _____.

f. The digit _____ is in the tens place. It has a value of _____.

g. The digit _____ is in the tenths place. It has a value of _____.

h. The digit _____ is in the hundredths place. It has a value of _____.

EUREKA MATH™

| Lesson 7: | Model mixed numbers with units of hundreds, tens, ones, tenths, and hundredths in expanded form and on the place value chart. |

3. Write each decimal as an equivalent fraction. Then, write each number in expanded form, using both decimal and fraction notation. The first one has been done for you.

Decimal and Fraction Form	Expanded Form	
	Fraction Notation	Decimal Notation
$14.23 = 14\frac{23}{100}$	$(1 \times 10) + (4 \times 1) + (2 \times \frac{1}{10}) + (3 \times \frac{1}{100})$ $10 \;+\; 4 \;+\; \frac{2}{10} \;+\; \frac{3}{100}$	$(1 \times 10) + (4 \times 1) + (2 \times 0.1) + (3 \times 0.01)$ $10 \;+\; 4 \;+\; 0.2 \;+\; 0.03$
$25.3 =$ _____		
$39.07 =$ _____		
$40.6 =$ _____		
$208.90 =$ _____		
$510.07 =$ _____		
$900.09 =$ _____		

EUREKA MATH™

Lesson 7:   Model mixed numbers with units of hundreds, tens, ones, tenths, and hundredths in expanded form and on the place value chart.

hundredths	
tenths	
.	
ones	
tens	
hundreds	

place value chart

Lesson 7:   Model mixed numbers with units of hundreds, tens, ones, tenths, and
             hundredths in expanded form and on the place value chart.

41

Name _____   Date _____

. Use the area model to represent $\frac{250}{100}$. Complete the number sentence.

a. $\frac{250}{100}$ = _____ tenths = _____ ones _____ tenths = __.____

b. In the space below, explain how you determined your answer to (a).

. Draw number disks to represent the following decompositions:

2 ones = _____ tenths

ones	.	tenths	hundredths

2 tenths = _____ hundredths

ones	.	tenths	hundredths

1 one 3 tenths = ____ tenths

ones	.	tenths	hundredths

2 tenths 3 hundredths = ____ hundredths

ones	.	tenths	hundredths

EUREKA MATH™ | Lesson 8:   Use understanding of fraction equivalence to investigate decimal
                                 numbers on the place value chart expressed in different units.

43

3. Decompose the units to represent each number as tenths.

   a. 1 = _____ tenths

   b. 2 = _____ tenths

   c. 1.7 = _____ tenths

   d. 2.9 = _____ tenths

   e. 10.7 = _____ tenths

   f. 20.9 = _____ tenths

4. Decompose the units to represent each number as hundredths.

   a. 1 = _____ hundredths

   b. 2 = _____ hundredths

   c. 1.7 = _____ hundredths

   d. 2.9 = _____ hundredths

   e. 10.7 = _____ hundredths

   f. 20.9 = _____ hundredths

5. Complete the chart. The first one has been done for you.

Decimal	Mixed Number	Tenths	Hundredths
2.1	$2\frac{1}{10}$	21 tenths $\frac{21}{10}$	210 hundredths $\frac{210}{100}$
4.2			
8.4			
10.2			
75.5			

EUREKA
MATH™

Lesson 8:   Use understanding of fraction equivalence to investigate decimal numbers on the place value chart expressed in different units.

Name _____   Date _____

. Use the area model to represent $\frac{220}{100}$. Complete the number sentence.

   a. $\frac{220}{100}$ = _____ tenths = _____ ones _____ tenths = ___.____

   b. In the space below, explain how you determined your answer to (a).

. Draw number disks to represent the following decompositions:

   3 ones = _____ tenths   3 tenths = _____ hundredths

ones	.	tenths	hundredths

ones	.	tenths	hundredths

   2 ones 3 tenths = ____ tenths   3 tenths 3 hundredths = ____ hundredths

ones	.	tenths	hundredths

ones	.	tenths	hundredths

EUREKA MATH™

| Lesson 8: | Use understanding of fraction equivalence to investigate decimal numbers on the place value chart expressed in different units. |

45

3. Decompose the units to represent each number as tenths.

    a. 1 = _____ tenths   b. 2 = _____ tenths

    c. 1.3 = _____ tenths   d. 2.6 = _____ tenths

    e. 10.3 = _____ tenths   f. 20.6 = _____ tenths

4. Decompose the units to represent each number as hundredths.

    a. 1 = _____ hundredths   b. 2 = _____ hundredths

    c. 1.3 = _____ hundredths   d. 2.6 = _____ hundredths

    e. 10.3 = _____ hundredths   f. 20.6 = _____ hundredths

5. Complete the chart. The first one has been done for you.

Decimal	Mixed Number	Tenths	Hundredths
4.1	$4\frac{1}{10}$	41 tenths $\frac{41}{10}$	410 hundredths $\frac{410}{100}$
5.3			
9.7			
10.9			
68.5			

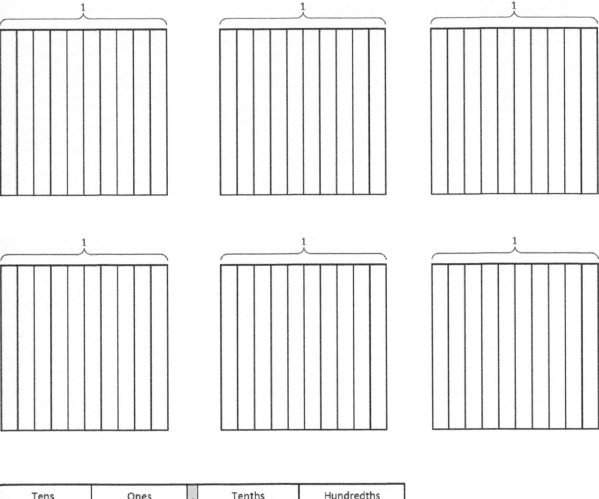

Tens	Ones	.	Tenths	Hundredths

area model and place value chart

Lesson 8:   Use understanding of fraction equivalence to investigate decimal
numbers on the place value chart expressed in different units.

47

Name _____   Date _____

Express the lengths of the shaded parts in decimal form. Write a sentence that compares the two lengths. Use the expression *shorter than* or *longer than* in your sentence.

a.

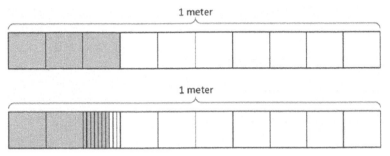

b.

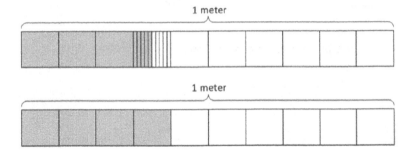

c. List all four lengths from least to greatest.

2. a. Examine the mass of each item as shown below on the 1-kilogram scales. Put an X over the items that are heavier than the avocado.

0.2 kg   0.12 kg   0.6 kg   0.61 kg

EUREKA
MATH™

Lesson 9: Use the place value chart and metric measurement to compare decimals and answer comparison questions.

49

b. Express the mass of each item on the place value chart.

	ones (kilograms) •	tenths	hundredths
avocado			
apple			
bananas			
grapes			

c. Complete the statements below using the words *heavier than* or *lighter than* in your statements.

The avocado is _____ the apple.

The bunch of bananas is _____ the bunch of grapes.

3. Record the volume of water in each cylinder on the place value chart below.

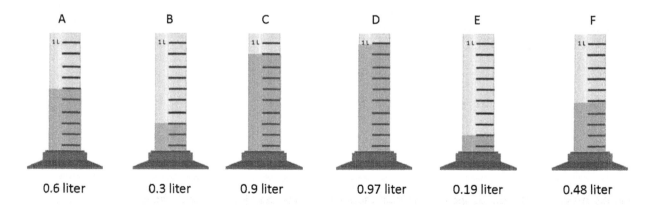

A	B	C	D	E	F
0.6 liter	0.3 liter	0.9 liter	0.97 liter	0.19 liter	0.48 liter

Cylinders	ones (Liters) .	tenths	hundredths
A			
B			
C			
D			
E			
F			

Compare the values using >, <, or =.

a. 0.9 L _____ 0.6 L

b. 0.48 L _____ 0.6 L

c. 0.3 L _____ 0.19 L

d. Write the volume of water in each beaker in order from least to greatest.

Name _____   Date _____

Express the lengths of the shaded parts in decimal form. Write a sentence that compares the two lengths. Use the expression *shorter than* or *longer than* in your sentence.

a.

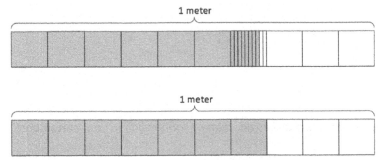

b.

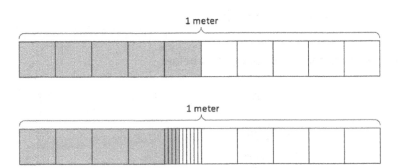

c.   List all four lengths from least to greatest.

a.   Examine the mass of each item as shown below on the 1-kilogram scales. Put an X over the items that are heavier than the volleyball.

0.15 kg   0.62 kg   0.43 kg   0.25 kg

Lesson 9:   Use the place value chart and metric measurement to compare decimals and answer comparison questions.

51

b. Express the mass of each item on the place value chart.

	ones (kilograms)	•	tenths	hundredths
baseball				
volleyball				
basketball				
soccer ball				

c. Complete the statements below using the words *heavier than* or *lighter than* in your statements.

The soccer ball is _____ the baseball.

The volleyball is _____ the basketball.

3. Record the volume of water in each cylinder on the place value chart below.

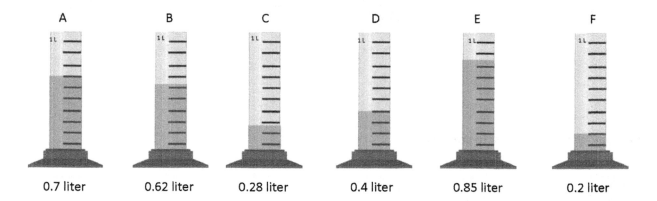

A	B	C	D	E	F
0.7 liter	0.62 liter	0.28 liter	0.4 liter	0.85 liter	0.2 liter

Cylinder	ones (liters)	.	tenths	hundredths
A				
B				
C				
D				
E				
F				

Compare the values using >, <, or =.

a. 0.4 L _____ 0.2 L

b. 0.62 L _____ 0.7 L

c. 0.2 L _____ 0.28 L

d. Write the volume of water in each beaker in order from least to greatest.

EUREKA MATH™

Lesson 9: Use the place value chart and metric measurement to compare decimals and answer comparison questions.
11/17/14

Rice Bag	ones (kilograms)	.	tenths	hundredths
A				
B				
C				
D				

Cylinder	ones (liters)	.	tenths	hundredths
A				
B				
C				
D				

measurement record

Lesson 9:   Use the place value chart and metric measurement to compare decimals and answer comparison questions.

53

Name _____ Date _____

. Shade the area models below, decomposing tenths as needed, to represent the pairs of decimal numbers. Fill in the blank with <, >, or = to compare the decimal numbers.

a. 0.23 _____ 0.4

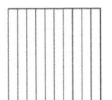 

b. 0.6 _____ 0.38

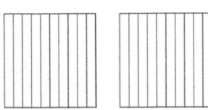

c. 0.09 _____ 0.9

 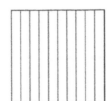

d. 0.70 _____ 0.7

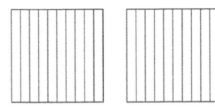

. Locate and label the points for each of the decimal numbers on the number line. Fill in the blank with <, >, or = to compare the decimal numbers.

a. 10.03 _____ 10.3

10.0   10.1   10.2   10.3

b. 12.68 _____ 12.8

12.6   12.7   12.8   12.9

Lesson 10:   Use area models and the number line to compare decimal numbers, and record comparisons using <, >, and =.

3. Use the symbols <, >, or = to compare.

    a. 3.42 _____ 3.75   b. 4.21 _____ 4.12

    c. 2.15 _____ 3.15   d. 4.04 _____ 6.02

    e. 12.7 _____ 12.70   f. 1.9 _____ 1.21

4. Use the symbols <, >, or = to compare. Use pictures as needed to solve.

    a. 23 tenths _____ 2.3   b. 1.04 _____ 1 one and 4 tenths

    c. 6.07 _____ $6\frac{7}{10}$ d. 0.45 _____ $\frac{45}{10}$

    e. $\frac{127}{100}$ _____ 1.72   f. 6 tenths _____ 66 hundredths

EUREKA
MATH™   Lesson 10:   Use area models and the number line to compare decimal numbers,
                           and record comparisons using <, >, and =.

Name _____ Date _____

Shade the parts of the area models below, decomposing tenths as needed, to represent the pairs of decimal numbers. Fill in the blank with <, >, or = to compare the decimal numbers.

a. 0.19 _____ 0.3

 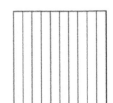

b. 0.6 _____ 0.06

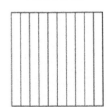 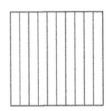

c. 1.8 _____ 1.53

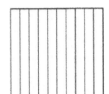 

d. 0.38 _____ 0.7

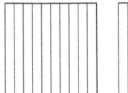 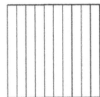

Locate and label the points for each of the decimal numbers on the number line. Fill in the blank with <, >, or = to compare the decimal numbers.

a. 7.2 _____ 7.02

7.0   7.1   7.2   7.3

b. 18.19 _____ 18.3

18.1   18.2   18.3   18.4

EUREKA MATH™

Lesson 10:   Use area models and the number line to compare decimal numbers, and record comparisons using <, >, and =.

57

3. Use the symbols <, >, or = to compare.

a. 2.68 _____ 2.54

b. 6.37 _____ 6.73

c. 9.28 _____ 7.28

d. 3.02 _____ 3.2

e. 13.1 _____ 13.10

f. 5.8 _____ 5.92

4. Use the symbols <, >, or = to compare. Use pictures as needed to solve.

a. 57 tenths _____ 5.7

b. 6.2 _____ 6 ones and 2 hundredths

c. 33 tenths _____ 33 hundredths

d. 8.39 _____ $8\frac{39}{10}$

e. $\frac{236}{100}$ _____ 2.36

f. 3 tenths _____ 22 hundredths

EUREKA
MATH™

Lesson 10: Use area models and the number line to compare decimal numbers, and record comparisons using <, >, and =.

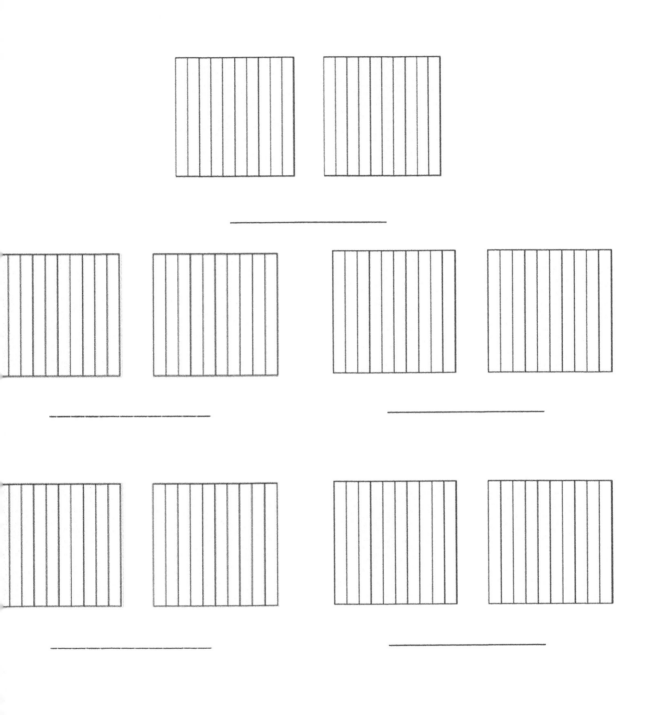

comparing with area models

Lesson 10:   Use area models and the number line to compare decimal numbers, and record comparisons using <, >, and =.

59

ame _____   Date _____

Plot the following points on the number line.

a. $0.2, \frac{1}{10}, 0.33, \frac{12}{100}, 0.21, \frac{32}{100}$

b. $3.62, 3.7, 3\frac{85}{100}, \frac{38}{10}, \frac{364}{100}$

c. $6\frac{3}{10}, 6.31, \frac{628}{100}, \frac{62}{10}, 6.43, 6.40$

EUREKA
MATH™

Lesson 11:   Compare and order mixed numbers in various forms.

61

2. Arrange the following numbers in order from greatest to least using decimal form. Use the > symbol between each number.

a. $\frac{27}{10}$, 2.07, $\frac{27}{100}$, $2\frac{71}{100}$, $\frac{227}{100}$, 2.72

b. $12\frac{3}{10}$, 13.2, $\frac{134}{100}$, 13.02, $12\frac{20}{100}$

c. $7\frac{34}{100}$, $7\frac{4}{10}$, $7\frac{3}{10}$, $\frac{750}{100}$, 75, 7.2

3. In the long jump event, Rhonda jumped 1.64 meters. Mary jumped $1\frac{6}{10}$ meters. Kerri jumped $\frac{94}{100}$ meter. Michelle jumped 1.06 meters. Who jumped the farthest?

4. In December, $2\frac{3}{10}$ feet of snow fell. In January, 2.14 feet of snow fell. In February, $2\frac{19}{100}$ feet of snow fell, and in March, $1\frac{1}{10}$ feet of snow fell. During which month did it snow the most? During which month did it snow the least?

EUREKA
MATH

Lesson 11: Compare and order mixed numbers in various forms.

Name _____   Date _____

.   Plot the following points on the number line using decimal form.

a. $0.6, \dfrac{5}{10}, 0.76, \dfrac{79}{100}, 0.53, \dfrac{67}{100}$

0.5   0.6   0.7   0.8

b.   8 ones and 15 hundredths, $\dfrac{832}{100}, 8\dfrac{27}{100}, \dfrac{82}{10}, 8.1$

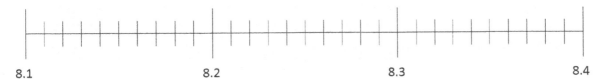

8.1   8.2   8.3   8.4

c. $13\dfrac{12}{100}, \dfrac{130}{10}, 13$ ones and 3 tenths, $13.21, 13\dfrac{3}{100}$

13.0   13.1   13.2   13.3

2. Arrange the following numbers in order from greatest to least using decimal form. Use the > symbol between each number.

   a. 4.03, 4 ones and 33 hundredths, $\frac{34}{100}$, $4\frac{43}{100}$, $\frac{430}{100}$, 4.31

   _____

   b. $17\frac{5}{10}$, 17.55, $\frac{157}{10}$, 17 ones and 5 hundredths, 15.71, $15\frac{75}{100}$

   _____

   c. 8 ones and 19 hundredths, $9\frac{8}{10}$, 81, $\frac{809}{100}$, 8.9, $8\frac{1}{10}$

   _____

3. In a paper airplane contest, Matt's airplane flew 9.14 meters. Jenna's airplane flew $9\frac{4}{10}$ meters. Ben's airplane flew $\frac{904}{100}$ meters. Leah's airplane flew 9.1 meters. Whose airplane flew the farthest?

4. Becky drank $1\frac{41}{100}$ liters of water on Monday, 1.14 liters on Tuesday, 1.04 liters on Wednesday, $\frac{11}{10}$ liters on Thursday, and $1\frac{40}{100}$ liters on Friday. Which day did Becky drink the most? Which day did Becky drink the least?

EUREKA MATH™

Lesson 11:   Compare and order mixed numbers in various forms.

3 tenths	0.2
0.17	$\dfrac{34}{100}$
13 hundredths	$\dfrac{4}{10}$

decimal number flash cards

Lesson 11:   Compare and order mixed numbers in various forms.

65

Name _____   Date _____

. Complete the number sentence by expressing each part using hundredths. Model using the place value chart, as shown in Part (a).

ones	•	tenths	hundredths

a. 1 tenth + 5 hundredths = _____ hundredths

ones	•	tenths	hundredths

b. 2 tenths + 1 hundredth = _____ hundredths

ones	•	tenths	hundredths

c. 1 tenth + 12 hundredths = _____ hundredths

. Solve by converting all addends to hundredths before solving.

a. 1 tenth + 3 hundredths = _____ hundredths + 3 hundredths = _____ hundredths

b. 5 tenths + 12 hundredths = _____ hundredths + _____ hundredths = _____ hundredths

c. 7 tenths + 27 hundredths = _____ hundredths + _____ hundredths = _____ hundredths

d. 37 hundredths + 7 tenths = _____ hundredths + _____ hundredths = _____ hundredths

Lesson 12: Apply understanding of fraction equivalence to add tenths and hundredths.

3. Find the sum. Convert tenths to hundredths as needed. Write your answer as a decimal.

   a. $\frac{2}{10} + \frac{8}{100}$ b. $\frac{13}{100} + \frac{4}{10}$

   c. $\frac{6}{10} + \frac{39}{100}$ d. $\frac{70}{100} + \frac{3}{10}$

4. Solve. Write your answer as a decimal.

   a. $\frac{9}{10} + \frac{42}{100}$ b. $\frac{70}{100} + \frac{5}{10}$

   c. $\frac{68}{100} + \frac{8}{10}$ d. $\frac{7}{10} + \frac{87}{100}$

5. Beaker A has $\frac{63}{100}$ liter of iodine. It is filled the rest of the way with water up to 1 liter. Beaker B has $\frac{4}{10}$ liter of iodine. It is filled the rest of the way with water up to 1 liter. If both beakers are emptied into a large beaker, how much iodine will be in the large beaker?

ame _____   Date _____

Complete the number sentence by expressing each part using hundredths. Model using the place value chart, as shown in Part (a).

ones		tenths	hundredths

a. 1 tenth + 8 hundredths = _____ hundredths

ones		tenths	hundredths

b. 2 tenths + 3 hundredths = _____ hundredths

ones		tenths	hundredths

c. 1 tenth + 14 hundredths = _____ hundredths

. Solve by converting all addends to hundredths before solving.

a. 1 tenth + 2 hundredths = _____ hundredths + 2 hundredths = _____ hundredths

b. 4 tenths + 11 hundredths = _____ hundredths + _____ hundredths = _____ hundredths

c. 8 tenths + 25 hundredths = _____ hundredths + _____ hundredths = _____ hundredths

d. 43 hundredths + 6 tenths = _____ hundredths + _____ hundredths = _____ hundredths

EUREKA
MATH™

Lesson 12:   Apply understanding of fraction equivalence to add tenths and
             hundredths.

69

3. Find the sum. Convert tenths to hundredths as needed. Write your answer as a decimal.

   a. $\frac{3}{10} + \frac{7}{100}$

   b. $\frac{16}{100} + \frac{5}{10}$

   c. $\frac{5}{10} + \frac{40}{100}$

   d. $\frac{20}{100} + \frac{8}{10}$

4. Solve. Write your answer as a decimal.

   a. $\frac{5}{10} + \frac{53}{100}$

   b. $\frac{27}{100} + \frac{8}{10}$

   c. $\frac{4}{10} + \frac{78}{100}$

   d. $\frac{98}{100} + \frac{7}{10}$

5. Cameron measured $\frac{65}{100}$ inch of rainwater on the first day of April. On the second day of April, he measured $\frac{83}{100}$ inch of rainwater. How many inches of rain fell on the first two days of April?

EUREKA
MATH™

Lesson 12:   Apply understanding of fraction equivalence to add tenths and
              hundredths.

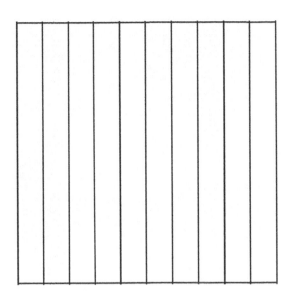

ones		tenths	hundredths
	●		

area model and place value chart

Lesson 12:   Apply understanding of fraction equivalence to add tenths and
hundredths.

71

ame _____   Date _____

Solve. Convert tenths to hundredths before finding the sum. Rewrite the complete number sentence in decimal form. Problems 1(a) and 1(b) are partially completed for you.

a. $2\frac{1}{10} + \frac{3}{100} = 2\frac{10}{100} + \frac{3}{100} =$ _____ $2.1 + 0.03 =$ _____	b. $2\frac{1}{10} + 5\frac{3}{100} = 2\frac{10}{100} + 5\frac{3}{100} =$ _____
c. $3\frac{24}{100} + \frac{7}{10}$	d. $3\frac{24}{100} + 8\frac{7}{10}$

. Solve. Then, rewrite the complete number sentence in decimal form.

a. $6\frac{9}{10} + 1\frac{10}{100}$	b. $9\frac{9}{10} + 2\frac{45}{100}$
c. $2\frac{4}{10} + 8\frac{90}{100}$	d. $6\frac{37}{100} + 7\frac{7}{10}$

3. Solve by rewriting the number sentence in fraction form. After solving, rewrite the complete number sentence in decimal form.

a. 6.4 + 5.3	b. 6.62 + 2.98
c. 2.1 + 0.94	d. 2.1 + 5.94
e. 5.7 + 4.92	f. 5.68 + 4.9
g. 4.8 + 3.27	h. 17.6 + 3.59

EUREKA
MATH™

Lesson 13:   Add decimal numbers by converting to fraction form.

Name _____ Date _____

Solve. Convert tenths to hundredths before finding the sum. Rewrite the complete number sentence in decimal form. Problems 1(a) and 1(b) are partially completed for you.

a. $5\frac{2}{10} + \frac{7}{100} = 5\frac{20}{100} + \frac{7}{100} = $ _____ $5.2 + 0.07 = $ _____	b. $5\frac{2}{10} + 3\frac{7}{100} = 8\frac{20}{100} + \frac{7}{100} = $ _____
c. $6\frac{5}{10} + \frac{1}{100}$	d. $6\frac{5}{10} + 7\frac{1}{100}$

Solve. Then, rewrite the complete number sentence in decimal form.

a. $4\frac{9}{10} + 5\frac{10}{100}$	b. $8\frac{7}{10} + 2\frac{65}{100}$
c. $7\frac{3}{10} + 6\frac{87}{100}$	d. $5\frac{48}{100} + 7\frac{8}{10}$

EUREKA
MATH™

Lesson 13: Add decimal numbers by converting to fraction form.

75

3. Solve by rewriting the number sentence in fraction form. After solving, rewrite the complete number sentence in decimal form.

a. $2.1 + 0.87 = 2\frac{1}{10} + \frac{87}{100}$	b. $7.2 + 2.67$
c. $7.3 + 1.8$	d. $7.3 + 1.86$
e. $6.07 + 3.93$	f. $6.87 + 3.9$
g. $8.6 + 4.67$	h. $18.62 + 14.7$

EUREKA
MATH™

Lesson 13:   Add decimal numbers by converting to fraction form.

Name _____   Date _____

Barrel A contains 2.7 liters of water. Barrel B contains 3.09 liters of water. Together, how much water do the two barrels contain?

Alissa ran a distance of 15.8 kilometers one week and 17.34 kilometers the following week. How far did she run in the two weeks?

Lesson 14:   Solve word problems involving the addition of measurements in decimal form.

77

3. An apple orchard sold 140.5 kilograms of apples in the morning and 15.85 kilograms more apples in the afternoon than in the morning. How many total kilograms of apples were sold that day?

4. A team of three ran a relay race. The final runner's time was the fastest, measuring 29.2 seconds. The middle runner's time was 1.89 seconds slower than the final runner's. The starting runner's time was 0.9 seconds slower than the middle runner's. What was the team's total time for the race?

Lesson 14:   Solve word problems involving the addition of measurements in decimal form.

ame _____   Date _____

The snowfall in Year 1 was 2.03 meters. The snowfall in Year 2 was 1.6 meters. How many total meters of snow fell in Years 1 and 2?

A deli sliced 22.6 kilograms of roast beef one week and 13.54 kilograms the next. How many total kilograms of roast beef did the deli slice in the two weeks?

Lesson 14:   Solve word problems involving the addition of measurements in
               decimal form.

79

3. The school cafeteria served 125.6 liters of milk on Monday and 5.34 more liters of milk on Tuesday than on Monday. How many total liters of milk were served on Monday and Tuesday?

4. Max, Maria, and Armen were a team in a relay race. Max ran his part in 17.3 seconds. Maria was 0.7 seconds slower than Max. Armen was 1.5 seconds slower than Maria. What was the total time for the team?

Lesson 14:   Solve word problems involving the addition of measurements in decimal form.

Name _____   Date _____

 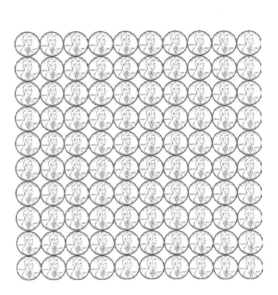

. 100 pennies = $___._____   100¢ = —— dollar
                                         100

. 1 penny = $___._____   1¢ = —— dollar
                                    100

. 6 pennies = $___._____   6¢ = —— dollar
                                     100

. 10 pennies = $___._____   10¢ = —— dollar
                                       100

. 26 pennies = $___._____   26¢ = —— dollar
                                       100

6. 10 dimes = $___._____   100¢ = —— dollar
                                            10

7. 1 dime = $___._____   10¢ = —— dollar
                                           10

8. 3 dimes = $___._____   30¢ = —— dollar
                                           10

9. 5 dimes = $___._____   50¢ = —— dollar
                                           10

10. 6 dimes = $___._____   60¢ = —— dollar
                                           10

1. 4 quarters = $___._____   100¢ = —— dollar
                                        100

2. 1 quarter = $___._____   25¢ = —— dollar
                                      100

3. 2 quarters = $___._____   50¢ = —— dollar
                                       100

4. 3 quarters = $___._____   75¢ = —— dollar
                                       100

EUREKA MATH | Lesson 15:   Express money amounts given in various forms as decimal numbers.

81

Solve. Give the total amount of money in fraction and decimal form.

15. 3 dimes and 8 pennies

16. 8 dimes and 23 pennies

17. 3 quarters, 3 dimes, and 5 pennies

18. 236 cents is what fraction of a dollar?

Solve. Express the answer as a decimal.

19. 2 dollars 17 pennies + 4 dollars 2 quarters

20. 3 dollars 8 dimes + 1 dollar 2 quarters 5 pennies

21. 9 dollars 9 dimes + 4 dollars 3 quarters 16 pennies

ame _____   Date _____

 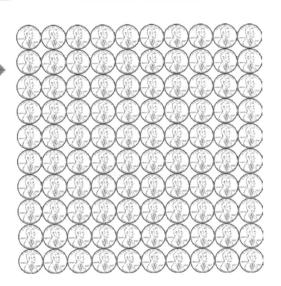

1. 100 pennies = $____._____   100¢ = $\frac{}{100}$ dollar

2. 1 penny = $____._____   1¢ = $\frac{}{100}$ dollar

3. 3 pennies = $____._____   3¢ = $\frac{}{100}$ dollar

4. 20 pennies = $____._____   20¢ = $\frac{}{100}$ dollar

5. 37 pennies = $____._____   37¢ = $\frac{}{100}$ dollar

6. 10 dimes = $____._____   100¢ = $\frac{}{10}$ dollar

7. 2 dimes = $____._____   20¢ = $\frac{}{10}$ dollar

8. 4 dimes = $____._____   40¢ = $\frac{}{10}$ dollar

9. 6 dimes = $____._____   60¢ = $\frac{}{10}$ dollar

10. 9 dimes = $____._____   90¢ = $\frac{}{10}$ dollar

11. 3 quarters = $____._____   75¢ = $\frac{}{100}$ dollar

12. 2 quarters = $____._____   50¢ = $\frac{}{100}$ dollar

13. 4 quarters = $____._____   100¢ = $\frac{}{100}$ dollar

14. 1 quarter = $____._____   25¢ = $\frac{}{100}$ dollar

EUREKA MATH™   Lesson 15:   Express money amounts given in various forms as decimal numbers.

83

Solve. Give the total amount of money in fraction and decimal form.

15. 5 dimes and 8 pennies

16. 3 quarters and 13 pennies

17. 3 quarters, 7 dimes, and 16 pennies

18. 187 cents is what fraction of a dollar?

Solve. Express the answer in decimal form.

19. 1 dollar 2 dimes 13 pennies + 2 dollars 3 quarters

20. 2 dollars 6 dimes + 2 dollars 2 quarters 16 pennies

21. 8 dollars 8 dimes + 7 dollars 1 quarter 8 dimes

EUREKA
MATH™

Lesson 15:   Express money amounts given in various forms as decimal numbers.

ame _____   Date _____

se the RDW process to solve. Write your answer as a decimal.

. Miguel had 1 dollar bill, 2 dimes, and 7 pennies. John had 2 dollar bills, 3 quarters, and 9 pennies. How much money did the two boys have in all?

. Suilin needed 7 dollars 13 cents to buy a book. In her wallet, she found 3 dollar bills, 4 dimes, and 14 pennies. How much more money does Suilin need to buy the book?

. Vanessa has 6 dimes and 2 pennies. Joachim has 1 dollar, 3 dimes, and 5 pennies. Jimmy has 5 dollars and 7 pennies. They want to put their money together to buy a game that costs $8.00. Do they have enough money to buy the game? If not, how much more money do they need?

4. A pen costs $2.29. A calculator costs 3 times as much as a pen. How much do a pen and a calculator cost together?

5. Krista has 7 dollars and 32 cents. Malory has 2 dollars and 4 cents. How much money does Krista need to give Malory so that each of them has the same amount of money?

Name _____   Date _____

Use the RDW process to solve. Write your answer as a decimal.

1. Maria had 2 dollars, 3 dimes, and 4 pennies. Lisa had 1 dollar and 5 quarters. How much money did the two girls have in all?

2. Meiling needed 5 dollars 35 cents to buy a ticket to a show. In her wallet, she found 2 dollar bills, 11 dimes, and 5 pennies. How much more money does Meiling need to buy the ticket?

3. Joe had 5 dimes and 4 pennies. Jamal had 2 dollars, 4 dimes, and 5 pennies. Jimmy had 6 dollars and 4 dimes. They wanted to put their money together to buy a book that costs $10.00. Did they have enough? If not, how much more did they need?

4. A package of mechanical pencils costs $4.99. A package of pens costs twice as much as a package of pencils. How much do a package of pens and a package of pencils cost together?

5. Carlos has 8 dollars and 48 cents. Alissa has 4 dollars and 14 cents. How much money does Carlos need to give Alissa so that each of them has the same amount of money?